Grades 5-8

FOCUS ON

MIDDLE SCHOOL

Laboratory Notebook

3rd Edition

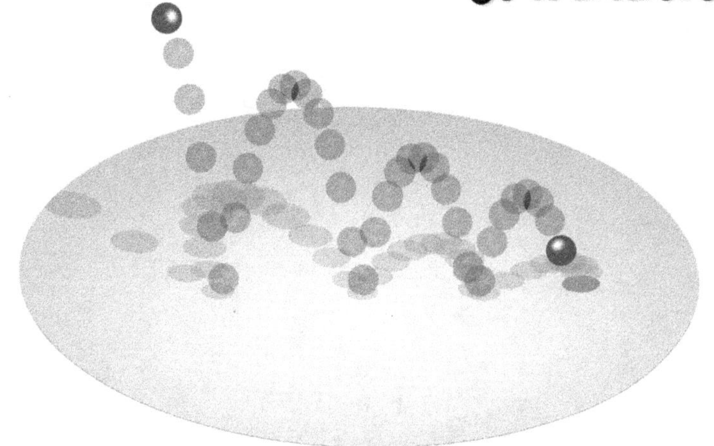

Rebecca W. Keller, PhD

Real Science-4-Kids

Illustrations: Janet Moneymake

Focus On Middle School Physics Laboratory Notebook—3rd Edition
ISBN 978-1-941181-74-4

Published by Gravitas Publications Inc.
www.gravitaspublications.com
www.realscience4kids.com

Keeping a Laboratory Notebook

A laboratory notebook is essential for the experimental scientist. In this type of notebook, the results of all your experiments are kept together along with comments and any additional information that is gathered. For this curriculum, you should use this book as your laboratory notebook and record your experimental observations and conclusions directly on its pages, just as a real scientist would.

The experimental section for each chapter is pre-written. The exact format of a notebook may vary among scientists, but all experiments written in a laboratory notebook have certain essential parts. For each experiment, a descriptive but short *Title* is written at the top of the page along with the *Date* the experiment is performed. Below the title, an *Objective* and a *Hypothesis* are written. The objective is a short statement that tells something about why you are doing the experiment, and the hypothesis is the predicted outcome. Next, a *Materials List* is written. The materials needed for the experiment should be gathered before the experiment is started.

Following the *Materials List* is the *Experiment*. The sequence of steps and all the details for performing the experiment are written beforehand. Any changes made during the experiment should be written down. Include all information that might be of some importance. For example, if you are to measure 237 ml (1 cup) of water for an experiment, but you actually measured 296 ml (1 1/4 cup), this should be recorded. It is hard sometimes to predict the way in which even small variations in an experiment will affect the outcome, and it is easier to track a problem if all of the information is recorded.

The next section is the *Results* section. Here you will record your experimental observations. It is extremely important that you be honest about what is observed. For example, if the experimental instructions say that a solution will turn yellow, but your solution turned blue, you must record blue. You may have done the experiment incorrectly, or you might have discovered a new and interesting result, but either way, it is very important that your observations be honestly recorded.

Finally, the *Conclusions* should be written. Here you will explain what the observations may mean. You should try to write only valid conclusions. It is important to learn to think about what the data actually show and also what cannot be concluded from the experiment.

Contents

Experiment 1

It's the Law!

Introduction

Use the scientific method to determine Newton's First Law of Motion!

I. Think About It

❶ When you drive a car, can you choose to follow the speed limit? Why or why not?

❷ Do you think laws like following the speed limit can be broken? Why or why not?

❸ When you throw a ball up, does it always come down? Why or why not?

❹ Do you think a ball will behave in the same way at the top of a mountain as it does at the bottom? Why or why not?

❺ Do you think a ball will behave in the same way on Earth and on the Moon? Why or why not?

❻ Are there physical laws that can be broken? Why or why not?

II. Experiment 1: It's the Law!

Date ___________

Objective In this experiment we will use the scientific method to determine Newton's First Law of Motion.

Hypothesis ____

____

____

Materials

tennis ball
yarn or string (3 meters [10 ft])
paperclip
marble
bouncing ball (1 or more)

EXPERIMENT

Part I

❶ Take the tennis ball outside and throw it as far as you can. Observe how the ball travels through the air. In the space below, sketch the path the ball traveled.

❷ Take the piece of string or yarn and attach it to the tennis ball as follows:

① Open the paperclip up on one side and make a hook on the end as shown below:

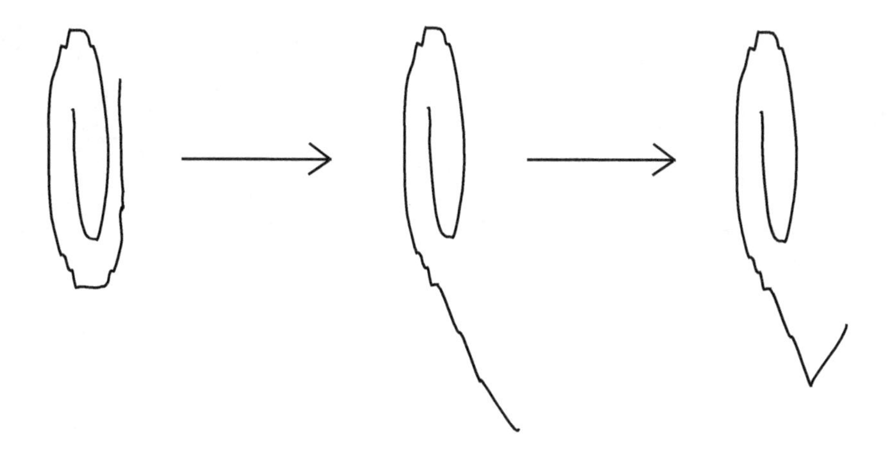

② Insert the hook on the end of the paperclip into the tennis ball by gently pushing and twisting.

③ Tie the string securely to the end of the paperclip.

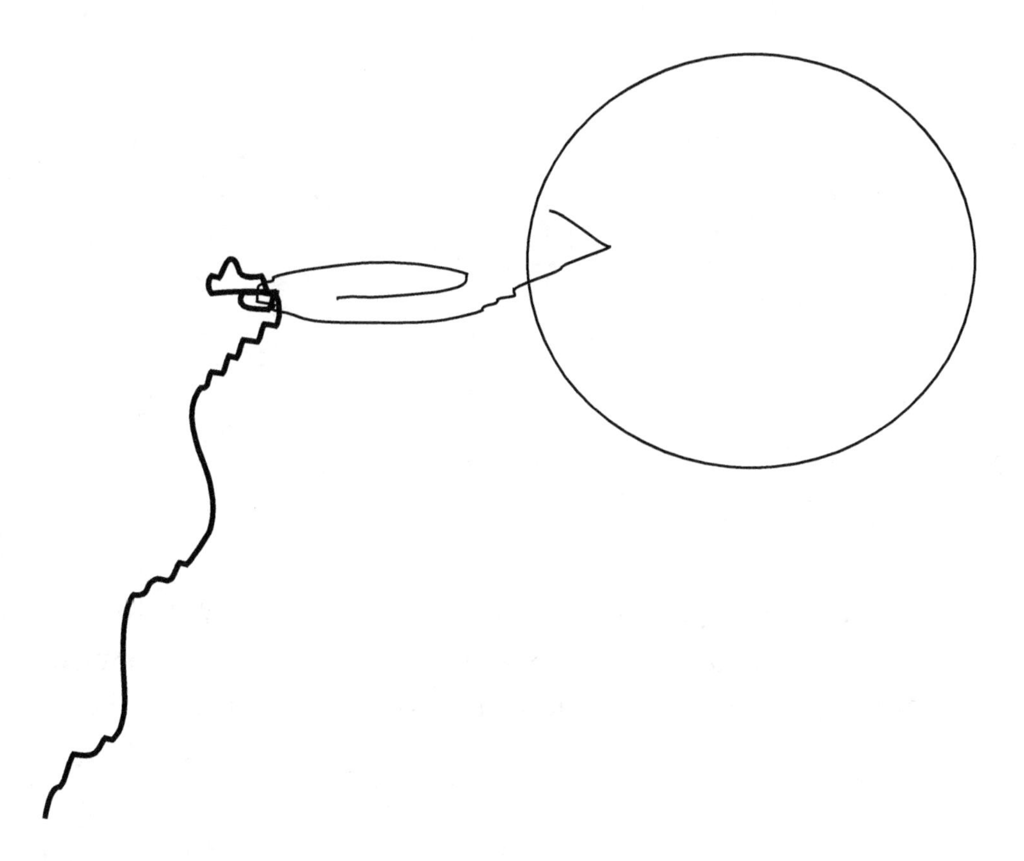

❸ Holding onto one end of the string, again throw the ball into the air as far as you can. Note how the ball travels, and in the space below, record what you see. Do this several times.

Part II

❶ Take a marble and find a straight, clear path on a smooth area of the floor or outdoors. Roll the marble and record how it travels. Note where and how it stops or changes direction. Do this several times and record your observations in the next box.

❷ Repeat Step ❶ using a rough surface on which to roll the marble.

III. Conclusions

Based on your observations, what conclusions can you draw from the results of this experiment?

IV. Why?

Physical laws are not laws we make up ourselves. The physical world is ordered, reliable, and consistent. This orderliness means there are underlying physical laws, or general principles, that we can discover to better understand the world. Physical laws are regularities that scientists have discovered in the way things behave. Physical laws are described by mathematics. Because the universe is ordered, mathematics can be used to precisely describe the laws that govern it.

In this experiment you discovered Newton's First Law of Motion by observing the movements of a tennis ball and a marble. Newton's First Law of Motion can be stated as: *A body will remain at rest or in motion until it is acted on by an outside force.*

By attaching one end of a long string to a tennis ball, you were able to observe a difference in how the ball traveled once it was thrown. The string changed the path the tennis ball followed. When the ball was thrown, it began traveling in an arc, but when the string reached its full length, the ball abruptly stopped and fell to the ground. The path the ball followed only changed when the string acted on it.

In a similar way, when the marble was rolled on a smooth surface, it traveled in a mostly straight line. When the marble was rolled on a rough surface, the irregularities of the surface changed the path traveled by the marble. If you roll a marble on a smooth surface and the marble runs into an obstacle such as small building block, you will observe the marble traveling straight until it contacts the obstacle. The obstacle provides the outside force to act on the marble and change the path it's following. The marble's trajectory, or path, is not changed unless it is contacted by something—like a rough surface or a building block.

V. Just For Fun

Play with a rubber bouncing ball. Bounce it softly. How high does it go? Bounce it hard. How high does it go this time?

How many times can you get it to bounce if you drop it softly? How many times can you get it to bounce if you drop it hard?

What happens if you bounce it using the same amount of force but vary the height from which you drop it?

If you have a bouncing ball of a different size, repeat the experiment and observe any differences.

Record your observations.

Experiment 2

Using Electronics

Introduction

Find out how electric circuits work!

I. Think About It

❶ How many modern toys, tools, and appliances can you think of that have electric circuits?

❷ What do you think an electric circuit looks like? Draw your idea here.

❸ What do you think the difference is between an electric circuit and an integrated circuit?

❹ Do you think integrated circuits include electric circuits? Why or why not?

❺ What do you think an integrated circuit looks like? Draw your idea here.

❻ How do you think toys, tools, and appliances with integrated circuits are different from those that have only electric circuits? Why?

II. Experiment 2: Using Electronics

Date ___________

Objective

____

____

Hypothesis

____

____

____

Materials

An electronic circuit kit

Recommended kits:

Snap Circuits: http://www.snapcircuits.net/
Snap Circuits Jr. 100 Experiments Kit

Little Bits: http://littlebits.cc/intro
Base Kit: http://littlebits.cc/kits/base-kit

EXPERIMENT

❶ Open the electronic kit and study the parts. Note whether you have switches, wires, a circuit grid, and/or clips.

❷ Read the instructions for using the parts and study any Do's and Don'ts for building electronic circuits.

❸ Start with the first project listed in the kit (For Snap Circuits Jr. 100 this is the Electric Light and Switch). Follow the directions and assemble the project. Record your observations. Include a diagram of the finished circuit, label the parts, and describe how it works. Note whether the project worked as described.

Observations—First Electronics Project

❹ Once you have built the first project, assemble the second project (For Snap Circuits Jr. 100 this is the DC Motor and Switch). Follow the directions and assemble the project. Record your observations. Include a diagram of the finished circuit, label the parts, and describe how it works. Note whether the project worked as described.

Results

Now that you have a basic understanding of the kit and circuits, work through several more projects and record your observations for each.

Observations: __________________ Project

Observations: __________________ Project

Observations: ________________ Project

Observations: ________________ Project

III. Conclusions

A. Questions

❶ How easy or difficult was it to build a motor, light switch, alarm, or other device?

____

____

____

____

❷ What is a closed circuit?

____

____

____

____

❸ What is a DC motor?

____

____

____

____

❹ What is resistance and how does it work?

____

____

____

____

❺ What is a fuse and what does it do?

B. Conclusions

Based on your observations, what conclusions can you draw from the circuits you built with the electronic circuit kit?

IV. Why?

In this experiment you built a variety of electric and integrated circuits by using an electronic circuit kit with an electrical grid. An electric circuit is a closed path through which electrons can flow. An integrated circuit is a combination of several electric circuits joined together along with components such as transistors, capacitors, and resistors. If you look inside many modern tools, toys, and appliances, you can find small circuit boards with sophisticated integrated circuits.

The invention of electric circuits led to profound changes in people's lives. The first battery that produced electrical current was made around 1800 by Alessandro Volta (1745-1827), an Italian physicist. In the US in the late 1800s Thomas Edison (1847-1931) experimented with electric circuits and designed the first coal-fired central power system that made it possible for people to have electric lighting at work and in their homes. The invention of many new appliances and machines was made possible by the availability of electricity and electric circuits. People no longer had to rely on candles or oil lamps for lighting or wood stoves for cooking food.

The addition of electronic components to electrical circuits to make integrated circuits has brought a huge variety of new technologies to the world. By knowing how electric circuits work, how to put them together, and how to make them convert electrical energy into light, sound, and motion, a growing number of tools are now available to scientists. Integrated circuits have made computers possible and are used in everything from programmable ovens to very advanced and complicated scientific instruments.

V. Just For Fun

Use the parts from your kit to invent your own circuit. Try combining a motor or fan, sound, light, and switches. What other parts can you include? What can you create?

Give your circuit a name and record your observations. Include a labeled diagram of your circuit and describe how it works.

Observations: ______________________________

Experiment 3

Fruit Works?

Introduction

Find out if fruit can do work!

I. Think About It

❶ Do you think if you and your brother or sister both carried the same box of books up the same flight of stairs, you would both be doing the same amount of work? Why or why not?

____

____

____

____

❷ Do you think you'd be doing the same amount of work if you carried a box of books up one flight of stairs and then carried a box of books weighing the same amount up two flights of stairs? Why or why not?

____

____

____

____

❸ If you and your friend each carried a box of books up the same flight of stairs, but your box weighted twice as much, would you both be doing the same amount of work? Why or why not?

____

____

____

____

❹ Do you think a piece of fruit can do work? Why or why not?

❺ If you wanted a piece of fruit to do work for you, what would you have it do? Why?

❻ If fruit can do work, do you think a watermelon could do more work than a lemon? Why or why not?

II. Experiment 3: Fruit Works?

Date ____________

Objective ____

____

Hypothesis ____

____

____

Materials

Slinky
several paperclips
1-2 apples
1-2 lemons or limes
1-2 oranges
1-2 bananas
spring balance scale or food scale
meterstick, yardstick, or tape measure
tape

EXPERIMENT

❶ Try to predict, just by "weighing" each piece of fruit in your hands, which piece of fruit will do the most work and which piece will do the least work on the spring that is in the scale.

❷ State your prediction as the hypothesis.

❸ Weigh each piece of fruit on the balance or food scale.

❹ Record the weights in the following chart.

Fruit	Weight (grams or ounces)

❺ Prepare the fruit for the experiment. Take a paperclip and stretch one side out to make a small hook like you did in Experiment 1. Place the hook in one of the pieces of fruit.

Repeat for each different kind of fruit you will be testing.

❻ Next, take the Slinky and hold it up to the level of your chest. Allow 10 to 15 coils to hang below your hand. You will have to hold most of the Slinky in your hand.

❼ Measure the distance from the floor to the bottom of the Slinky with the meterstick, yardstick, or tape measure. Record your result below.

Distance from floor to Slinky with no fruit attached

❽ Take a piece of fruit with a hook in it and attach it to the end of the Slinky. Hold the Slinky at the same height as in Step ❻ with the same number of coils hanging below your hand. Allow the Slinky to be pulled down by the fruit.

❾ Use the meterstick, yardstick, or tape measure to measure from the end of the Slinky to the floor. Record your results in the following chart in the *Distance Floor to Slinky (With Fruit)* column.

Fruit	Distance Floor to Slinky (With Fruit)	Distance Floor to Slinky (No Fruit)	Distance Extended

⑩ Repeat Steps ❽ and ❾ with different kinds of fruit. Record your results each time in the following chart.

Results

❶ Using the above chart, in each row of the *Distance Floor to Slinky (No Fruit)* column, write the distance you recorded in Step ❼. Then subtract this distance from each of the distances you recorded in the *Distance Floor to Slinky (With Fruit)* column. This gives you the distance the Slinky was extended by each piece of fruit.

❷ Using the formula *work= distance x force* where force is the weight of the fruit, calculate the work each piece of fruit has done. Record your answers in the following chart.

Fruit	Work

III. Conclusions

What conclusions can you draw from your observations?

IV. Why?

The concept of work may be difficult to understand because when we hear the word "work," we think of mowing the lawn or doing the laundry. However, in physics, work is defined as:

work = distance x force

By definition, a force is something that changes the position, shape, or speed of an object.

The illustration in the *Student Textbook* shows that, for the same amount of force, the work a short weight lifter does is less than the work a tall weight lifter does because the distance of the lift is less for the short weight lifter.

In this experiment, the heavier the fruit the more it stretched out the Slinky. The distance the fruit traveled was greater, so more work was done.

Another example: If you carry a box of books up one flight of stairs, and your brother carries the same box up two flights of stairs, who has done more work? Your brother because the books were carried a greater distance. In fact, your brother has done twice as much work because he carried the books twice as far.

If you carry a box of books up one flight of stairs, and your brother carries a box of books that weigh half as much up the same flight of stairs, you have done more work than your brother. In fact, you have done twice as much work.

In this experiment, by using the formula *work = distance x force* you found that the fruit that weighed the most did the most work.

V. Just For Fun

❶ What would happen if you attached two pieces of the same kind of fruit to the Slinky? How much work would be done?

Prediction ______________________________

____

____

____

2 Pieces of Fruit	Weight	Distance: Floor to Slinky (With Fruit)	Distance Floor to Slinky (No Fruit)	Distance Extended

❷ Test your prediction and calculate the work that was done by two pieces of fruit together. Record your data in the charts below.

Repeat this step one or more times using 2 pieces of fruit.

2 Pieces of Fruit	Work

Experiment 4

Smashed Banana

Introduction

Do you think a toy car can do work on a banana by gravitational potential energy being converted to kinetic energy? Try this experiment!

I. Think About It

❶ Do you think gravitational potential energy is useful? Why or why not?

____

____

____

____

____

❷ What kinds of things can kinetic energy be used for?

____

____

____

____

____

❸ Do you think you use gravitational potential energy every day? If so, how?

____

____

____

____

____

❹ Do you think you use kinetic energy every day? If so, how?

❺ What do you think would happen if gravitational potential energy could not be converted?

❻ What would happen if there were no kinetic energy?

II. Experiment 4: Smashed Banana

Date ___________

Objective

____

____

Hypothesis

____

____

____

Materials

small to medium size toy car
stiff cardboard
wooden board (more than 1 meter [3 feet] long)
straight pin or tack, several
small scale or balance
1 banana, sliced
10 pennies
meterstick, yardstick, or tape measure
tape

EXPERIMENT

❶ Read through all the steps of this experiment. Then write an objective and a hypothesis.

❷ Take a portion of the cardboard to make a backing for the banana slices. Using straight pins or tacks, attach two or three banana slices next to each other on the cardboard near the bottom.

❸ Use the wooden board to make a ramp. One end of the ramp should meet the banana slices. Your setup should look like the following illustration.

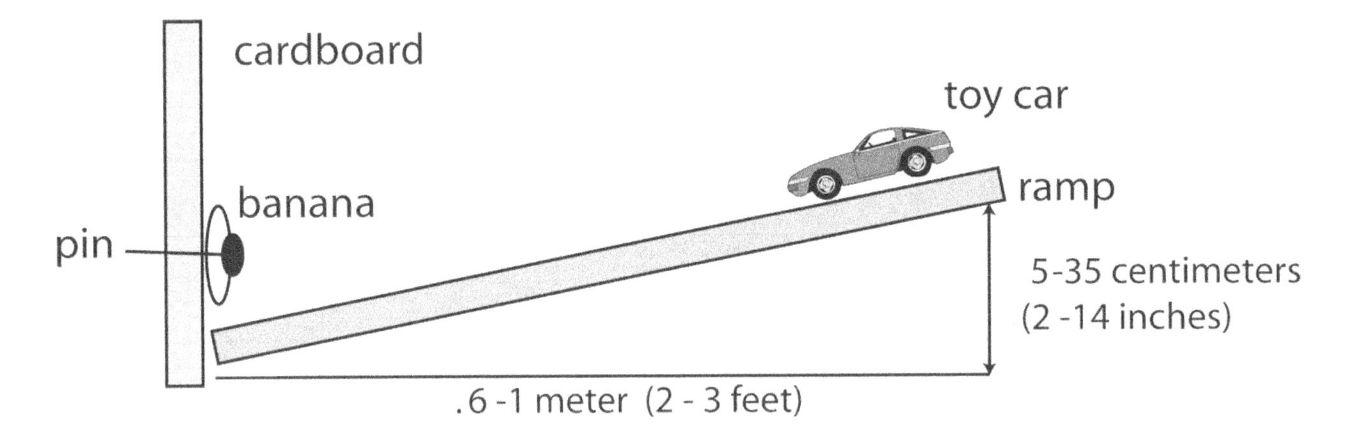

❹ Weigh the toy car with the scale or balance. Record your result.

Weight of toy car (grams or ounces) = ______________________________

❺ Place the toy car on the ramp and elevate one end of the ramp 5 centimeters (2 inches). Allow the toy car to roll down the ramp and hit the banana. Record your results in the following chart.

❻ Elevate the ramp another 5 centimeters (2 inches). Now the ramp should be 10 centimeters (4 inches) off the ground. Allow the toy car to roll down the ramp and hit the banana. Record your results in the chart below

❼ Repeat, elevating the ramp 5 centimeters (2 inches) more each time. Record your results in the following chart.

Height (centimeters or inches)	Results (write your comments)
5 centimeters (2 inches)	
10 centimeters (4 inches)	
15 centimeters (6 inches)	
20 centimeters (8 inches)	
25 centimeters (10 inches)	
30 centimeters (12 inches)	
35 centimeters (14 inches)	

❽ What happened to the speed of the car as the ramp height increased?

____

At which ramp height did the car smash the banana?

____

❾ Now add 10 pennies to the toy car and weigh it again. Repeat the previous steps, rolling the toy car with the pennies on it down the ramp and elevating the ramp 5 centimeters (2 inches) more each time Record your results.

Weight of toy car plus 10 pennies (grams or ounces) = ____________________

Height (centimeters or inches)	Results (write your comments)
5 centimeters (2 inches)	
10 centimeters (4 inches)	
15 centimeters (6 inches)	
20 centimeters (8 inches)	
25 centimeters (10 inches)	
30 centimeters (12 inches)	
35 centimeters (14 inches)	

❿ At which ramp height did the car smash the banana? ____________________

Was the banana smashed at the same height by the light car and the heavy car?

____

If "no," which car needed to be at a greater height to smash the banana?

____

Results

Calculate the GPE for the height of the ramp at which the toy car — with and without the 10 pennies — smashed the banana. Use the equation:

gravitational potential energy (GPE) = weight x height

Record your answers below.

GPE for car without pennies ______________________________

GPE for car with pennies ______________________________

Is the GPE the same (or close to the same) for both cars? ______________

III. Conclusions

What conclusions can you draw from your observations?

IV. Why?

Energy exists in different forms and is converted from one form to another. In this experiment you used the conversion of two different types of energy — potential energy and kinetic energy — to do work when the toy car smashed a banana. Potential energy is energy that has the potential to do work, and kinetic energy is the energy of motion. The gravitational potential energy (GPE) of the toy car on the elevated ramp was converted into kinetic energy (KE) as the toy car moved down the ramp.

The amount of GPE an object has is equal to the amount of work that was needed to lift the object to its current position. Each time you raised the ramp, you added more GPE to the toy car. The higher you lifted the ramp, the more GPE the toy car had and the more KE it had as the GPE was converted

Kinetic energy is proportional to both the weight of the object and its speed. Heavier objects will have more KE at a given speed than lighter objects, and slower objects will have less KE at a given weight than faster objects. When you put pennies on the toy car, it was heavier and moved faster and so gained KE.

Recall that work is the force of an object multiplied by the distance the object is moved. We know that an object having kinetic energy is moving, and if it hits another object, it can cause the object it hits to move or change shape. There is work done when potential energy is converted into kinetic energy and when kinetic energy is converted into other forms of energy, such as heat and sound. The work done on an object equals the change in kinetic energy of that object. When the moving toy car contacted the banana, the kinetic energy was converted into other forms of energy and work was done to the banana.

Potential energy is useful only when it gets converted to another form of energy. Energy is neither created nor destroyed—but only converted from one form to anther.

V. Just For Fun

What happens if you replace the banana with a raw egg? Do you need more or less gravitational force to smash an egg than a banana? Why?

Record your observations in the following space.

Observations of a Smashed Egg

Experiment 5

On Your Own

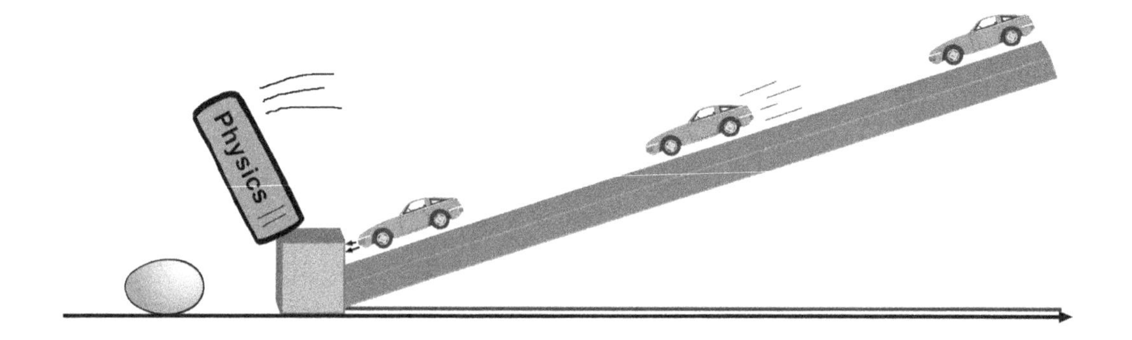

Introduction

Create your own experiment to explore the conversion of energy.

I. Think About It

1. Do you think potential energy can be used without converting it to another form of energy? Why or why not?

2. Do you think an object could go back and forth between having gravitational potential energy and kinetic energy? Why or why not?

3. Do you think you could gather energy from the Sun and use it? Why or why not?

❹ Do you think you use energy from chemical reactions during the day? Why or why not?

❺ What do you think life would be like if we did not have fossil fuels?

❻ Do you think water can have energy? Why or why not?

II. Experiment 5: On Your Own

This time you get to design your own experiment. The goal is to convert as many forms of energy as you can into other forms of energy. You can use any type of energy conversion more than once.

Example:

A scenario can be designed in which energy is used to put out a fire. A marble is rolled down a ramp and bumps into a domino that has on top of it a small cap containing baking soda. A chemical reaction is started when the baking soda falls into vinegar, which produces carbon dioxide gas that puts out the fire.

In this case, the marble begins with gravitational potential energy that is converted to kinetic energy as the marble rolls. The kinetic energy of the rolling marble is used to convert gravitational potential energy into kinetic energy (the falling baking soda) which then starts a chemical reaction.

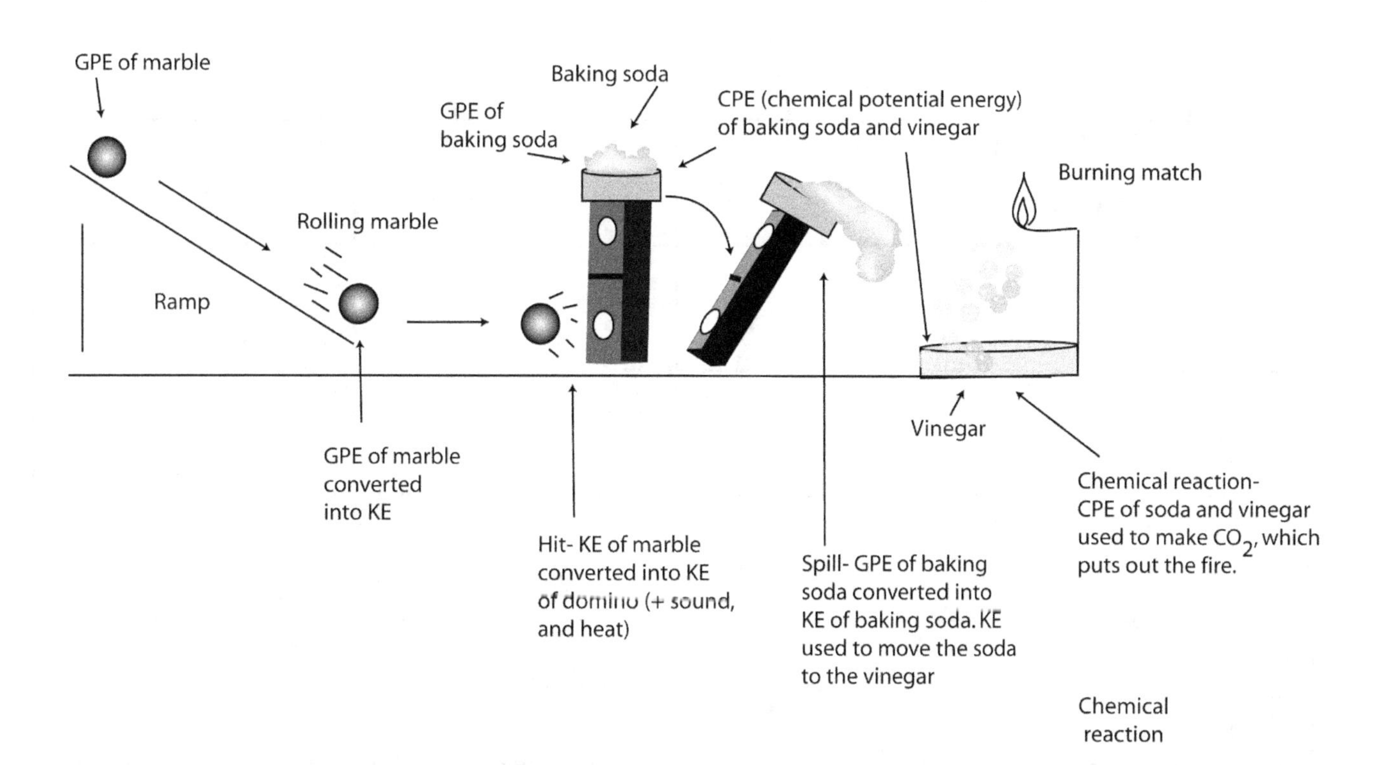

Use the following guide to design your experiment:

❶ Write down as many different forms of energy as you can think of.

kinetic energy

❷ Write down different ways each of these forms of energy can be represented.

Form of Energy	Represented by	Represented by	Represented by
kinetic energy	*rolling marble*	*moving toy car*	*moving ball*

❸ Write down different ways to connect two or more of these forms of energy and explain how one form will be converted into another.

moving toy car bumps into marble and starts it rolling

(kinetic energy converts potential energy into kinetic energy)

❹ Design an experiment to convert energy from one form into another. Give your experiment a title and write an objective and a hypothesis. Write down the materials you will need, and then write down the steps you will take to collect the results. Draw your setup. See how many different forms of energy you can convert.

❺ Perform the experiment, making and recording careful observations. Then draw conclusions based on what you observe. Use extra paper if needed.

Experiment: ______________________ Date __________

Objective

Hypothesis

Materials

EXPERIMENT

III. Conclusions

What conclusions can you draw from your observations?

IV. Why?

The law of conservation of energy states that energy is never created or destroyed but only converted from one form to another. In this experiment you took several different types of energy, and by connecting them, converted energy from one form to another. For example, if you used a marble to knock down a popsicle stick covered in baking soda and the baking soda then fell into a thimble full of vinegar, you converted kinetic and potential energy into chemical and heat energy. By converting energy from one form to another, you showed the law of conservation of energy in action.

V. Just For Fun

Based on what you learned in your experiment, create another experiment with a different series of events that convert energy. How many different steps can you include in your setup? List them below and draw your experimental setup on the next page. Once you have completed your setup, perform the experiment.

Energy Conversion Steps

Experimental Setup

III. Conclusions

What conclusions can you draw from your observations of this experiment? Compare your two experiments. Was one easier to perform? Why or why not? Did you learn things by performing the first experiment that helped you plan the second one? Why or why not?

Experiment 6

Moving Marbles

Introduction

What happens when different size rolling marbles meet?

I. Think About It

❶ Do you think friction can affect a rolling marble? Why or why not?

❷ Do you think a heavy marble could roll differently than a light marble? Why or why not?

❸ Do you think it might be harder to start a tennis ball rolling than a bowling ball? Why or why not?

❹ Do you think inertia is involved when you hit a baseball with a bat? Why or why not?

____

____

____

____

____

❺ Do you think it would be easier to catch a baseball that has more momentum or one that has less momentum? Why

____

____

____

____

____

❻ What do you think would happen if there were no friction? Why?

____

____

____

____

____

II. Experiment 6: Moving Marbles

Date ____________

Objective

____

____

Hypothesis

____

____

____

Materials

several glass marbles of different sizes
several steel marbles of different sizes
cardboard tube, .7-1 meter long (2 1/2-3 ft)
scissors
black marking pen
ruler
letter scale or other small scale or balance

EXPERIMENT

❶ Using the scale, weigh each of the marbles, both glass and steel. Keep track of how much each marble weighs by using the marking pen to label the marbles with numbers or letters, or you can note their colors. Record the information for each marble in *Part A* of the *Results* section.

❷ Take the cardboard tube and cut it in half lengthwise to make a trough. Measure the length of the cardboard trough, and mark the halfway point with the black marking pen.

❸ Beginning at the halfway mark, measure .3 meter (1 foot) in both directions, and put a mark at each of these measurements. This will give you one mark on each side of the halfway mark.

❹ The cardboard trough should now have three marks: one at the halfway point, and one on either side of the halfway mark, .3 meter (1 foot) away from it. The trough will be used as a track for the marbles.

❺ Take the marbles and, one by one, roll them down the trough. Notice how each one rolls (Does it roll straight? Is it easy to push off with your thumb? Does it pass the marks you drew?) In *Part B* of the *Results* section, describe how each marble rolls.

❻ Now place a glass marble on the center mark of the trough.

❼ Roll a glass marble of the same size toward the marble in the center. Watch the two marbles as they collide. Record your results in the *Results* section, *Part C*.

❽ Repeat Steps ❻ and ❼ with different size marbles. For example, try rolling a heavy marble toward a light marble and a light marble toward a heavy marble. Record your results in *Part D*.

Results

Part A—Marble Descriptions

Marble	Weight

Part B—How Marbles Roll

Part C—Same Size Colliding Marbles

Part D—Different Size Colliding Marbles

Other Observations

III. Conclusions

Review your observations. What conclusions can you draw from them?

IV. Why?

In this experiment you observed how the forces of *inertia, momentum,* and *friction* affected rolling and stationary marbles.

Inertia is the tendency of things to resist a change in motion. In physics there are two aspects to consider with regard to inertia: *mass* and *momentum*.

It is important to understand that *mass* and *weight* are different. Weight is a force. Mass is not. However, by weighing an object you can tell how much mass it has. The more mass an object has, the more it will weigh on Earth because gravity will exert more force on an object with a lot of mass than on an object with less mass. Without gravity, objects do not weigh anything. In space, where there is no gravity, a boulder would float in the same way a feather would. However, in space, the boulder and the feather would still have different masses. The boulder would still have more mass than the feather and, as a result, would still be harder to accelerate (speed up) than a feather — even in space.

The second aspect of inertia is the fact that an object with a lot of *momentum* is hard to stop. Momentum is inertia in motion; that is, mass that is moving. An object that has a large mass will have a large momentum. Also, an object moving at a fast rate of speed will have a large momentum. The mathematical equation for momentum is:

$$momentum = mass \times speed$$

Recall Newton's First Law of Motion that states: An object in motion will stay in motion unless acted on by an outside force, and an object at rest will stay at rest unless acted on by an outside force. Because inertia is the tendency of an object to resist a change in motion, objects that are stationary want to remain stationary, and objects that are moving want to remain in motion.

Although inertia keeps things moving, objects on Earth will eventually stop because a force acts on the object. This force is *friction*. Friction occurs when two objects rub against each other. Friction is the force that works in the direction opposite the direction of motion. Friction is what slows objects down and eventually causes them to stop. In the absence of friction, an object would keep moving forever and never stop. In this experiment, there is friction between the cardboard tube and the marbles.

V. Just For Fun

Repeat the experiment, this time using a baseball, a basketball, and a golf ball. Think about whether you might have to modify the experimental setup. If you do want to modify it, what would you change and how would you change it?

Record your results. Draw conclusions about how differences in mass, inertia, momentum, speed, and friction affect this experiment.

Descriptions	
Object Name and Weight	**How It Rolls**

Collision Descriptions

Conclusions

Review your observations. What conclusions can you draw about how differences in mass, inertia, momentum, speed, and friction affected this experiment?

Experiment 7

Accelerate to Win!

Introduction

How can knowing about velocity and acceleration help you win a race?

I. Think About It

❶ How do you think long distance runners win a race?

❷ What do you think happens on the last lap of the Indy 500?

❸ If you were riding in a horse race, what do you think you might do to win?

❹ What do you think you would do if you were neck-and-neck with your best friend in a running race?

____

____

____

____

____

❺ How do you think you could win a bike race with friends?

____

____

____

____

____

❻ How do you think you might train for winning a foot race? What do you think you would need to know?

____

____

____

____

____

____

____

____

II. Experiment 7: Accelerate to Win! Date ___________

Objective ____

____

Hypothesis ____

____

____

Materials

stopwatch
compass
an open space large enough to run (park, schoolyard, playground, backyard, etc.)
5 markers of your choice to mark distances
a group of friends

EXPERIMENT

Imagine that you are training for the final race of an Olympic running race and you are determined to win. You have to go the full distance without stopping before the end and you need to go fast enough to win. You are going to follow your coach's recommendation and start slowly and then sprint as fast as you can for the last quarter of the race.

❶ Map out a straight "track" and mark a starting and stopping point.

❷ Take the compass and find out the direction you will be running in. Record this direction on the chart in the *Results* section.

❸ Measure the distance between the starting point and the stopping point by walking heel-to-toe and counting each step as one "foot." Record the distance here.

Length of track in "feet" ______________________________

4. Take your measurement of the length of the track and divide it into fourths. Record the distances to the points at which you will time your run. Each time point distance is measured from the previous time point.

   Time point 1: $\mathbf{d_1}$ (1/4 mark) ____________________

   Time point 2: $\mathbf{d_2}$ (1/2 mark) ____________________

   Time point 3: $\mathbf{d_3}$ (3/4 mark) ____________________

   Time point 4: $\mathbf{d_4}$ (Finish) ____________________

   Now record distances $\mathbf{d_1}$-$\mathbf{d_4}$ on the chart in the Results section.

5. On the track, measure with your feet the distance between each time point and mark time points $\mathbf{d_1}$, $\mathbf{d_2}$, and $\mathbf{d_3}$. [$\mathbf{d_4}$ (Finish) is already marked.]

6. Pick one person to run the stopwatch. Have a second person use the chart in the *Results* section to record your time at each of the time points.

7. Get ready! **Set! GO!**

8. Repeat three or four times or until you are too tired to continue.

Results

1. For each trial, use the formulas provided in the following chart to calculate the velocity at each time point. Space is provided for calculations. Record your results in the chart.

2. For each trial, use the formulas provided in the chart to calculate the acceleration between each time point. Record your results in the chart.

Time Trial Results

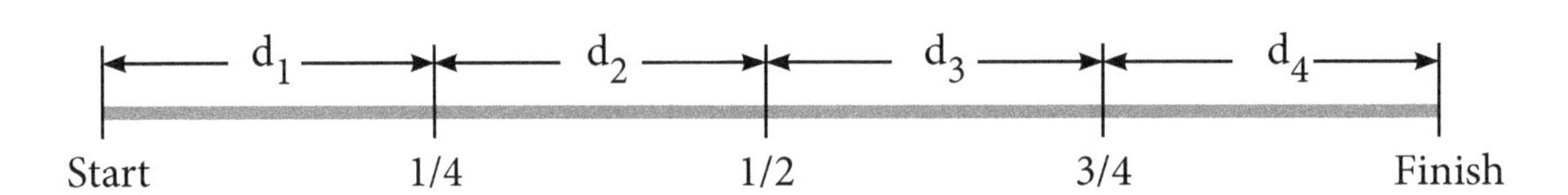

Direction ____________________

Distance (in "feet")

d_1 __________ d_2 __________ d_3 __________ d_4 __________

Time (seconds)

	t_1	t_2	t_3	t_4
Trial 1	________	________	________	________
Trial 2	________	________	________	________
Trial 3	________	________	________	________

Velocity

	$v_1 = \frac{d_1}{t_1}$	$v_2 = \frac{d_2}{t_2}$	$v_3 = \frac{d_3}{t_3}$	$v_4 = \frac{d_4}{t_4}$
Trial 1	________	________	________	________
Trial 2	________	________	________	________
Trial 3	________	________	________	________

Acceleration*

	$\mathbf{a}_1 = \left(\frac{v_2 - v_1}{\lvert t_2 - t_1 \rvert}\right)$	$\mathbf{a}_2 = \left(\frac{v_3 - v_2}{\lvert t_3 - t_2 \rvert}\right)$	$\mathbf{a}_3 = \left(\frac{v_4 - v_3}{\lvert t_4 - t_3 \rvert}\right)$
Trial 1	________	________	________
Trial 2	________	________	________
Trial 3	________	________	________

***Note:** For acceleration, time is always a positive number. In the acceleration formula, the change in time (Δt) is written as $\lvert t_f - t_i \rvert$ to show that the result is expressed as a positive number.

A Place for Your Calculations

III. Conclusions

A. Questions

❶ Which segment did you run with the fastest velocity? Why?

❷ Which segment did you run with the slowest velocity? Why?

❸ What can you notice about your acceleration in the different trials?

❹ In how many segments was your acceleration positive? negative? Which ones? Why?

B. Conclusions

Compare your trials. How was your performance in each? Were you faster or slower on the third trial? Explain your observations and results.

IV. Why?

By measuring the time it takes you to run between different points of known distance, you can calculate your velocity and acceleration. If you were training for the Olympics, by knowing how much energy you have and how fast you can go for how long, you could monitor how well you are doing in each run. You might notice that if you start out the run with a fast pace and accelerate too much near the beginning of the race, you are likely to run out of energy and slow down, decelerating near the finish line. Running out of energy before the end of the race won't help you to win, but you can learn how to start more slowly, run at a steady pace, and then accelerate at the finish.

V. Just For Fun

Run every day for a few weeks, recording the date and length of time. Then repeat the experiment. Record your results in the following chart. Have your times improved? What changes have you made in how you run a race? Use additional paper for observations such as the route you follow, weather, etc.

Date and Length of Time for Each Run

Date	Time	Date	Time	Date	Time

Time Trial Results

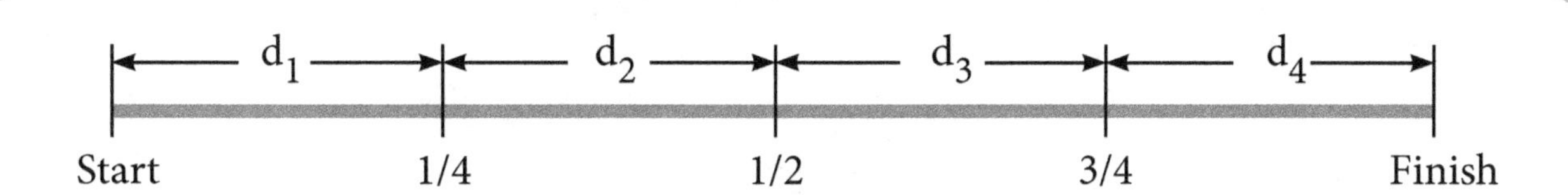

Direction ____________________

Distance (in "feet")

d_1 __________ d_2 __________ d_3 __________ d_4 __________

Time (seconds)	t_1	t_2	t_3	t_4
Trial 1	__________	__________	__________	__________
Trial 2	__________	__________	__________	__________
Trial 3	__________	__________	__________	__________

Velocity	$v_1 = \frac{d_1}{t_1}$	$v_2 = \frac{d_2}{t_2}$	$v_3 = \frac{d_3}{t_3}$	$v_4 = \frac{d_4}{t_4}$
Trial 1	__________	__________	__________	__________
Trial 2	__________	__________	__________	__________
Trial 3	__________	__________	__________	__________

Acceleration*	$a_1 = \left(\frac{v_2 - v_1}{\lvert t_2 - t_1 \rvert}\right)$	$a_2 = \left(\frac{v_3 - v_2}{\lvert t_3 - t_2 \rvert}\right)$	$a_3 = \left(\frac{v_4 - v_3}{\lvert t_4 - t_3 \rvert}\right)$
Trial 1	__________	__________	__________
Trial 2	__________	__________	__________
Trial 3	__________	__________	__________

***Note:** For acceleration, time is always a positive number. In the acceleration formula, the change in time (Δt) is written as $\lvert t_f - t_i \rvert$ to show that the result is expressed as a positive number.

There is space for calculations on the following page.

A Place for Your Calculations

Experiment 8

Around and Around

Introduction

Measure tangential speed!

I. Think About It

❶ If you run around your backyard in a circle, how far do you go?

❷ If you run around a circular running track how far do you go?

❸ If you could run around the Earth, how far would you go?

❹ Which is bigger, the diameter of a circle you can run in your backyard or the diameter of the Earth?

❺ If you could complete one lap around the Earth in the same length of time that you could complete one lap around your backyard, would you be running faster or slower in your trip around the Earth? Why?

____

____

____

____

____

❻ Do you think tangential speed is the same at any location on a disk that is spinning? Why or why not?

____

____

____

____

____

❼ Do you think tangential speed and rotational speed are related? Why or why not?

____

____

____

____

____

II. Experiment 8: Around and Around

Date ___________

Objective

____

____

Hypothesis

____

____

Materials

pen or pencil
marking pen
thumbtack or pushpin
3 pieces of string —
  approximate sizes:
  10 cm [4 in.]; 15 cm [6 in.]; 20 cm [8 in.]
tape
ruler
large piece of white paper (bigger than 30 cm [12 in.] square)

EXPERIMENT

❶ Lay the white sheet of paper on a flat, firm surface and use the thumbtack or pushpin to pin one end of the shortest string to the center of the paper.

❷ Measure the string beginning at the thumbtack and put a mark at 5 cm (2 in.).

❸ Take the pen or pencil and place it at the 5 cm mark on the string. Then wrap the extra length of string around the pen or pencil and fasten it with tape.

❹ Holding the thumbtack down, move the pen away from the thumbtack until the string is stretched out.

❺ Place the point of the pen in contact with the paper, holding the pen in a perpendicular position. Draw a circle around the center point.

❻ Repeat Steps ❶-❺ with the other two pieces of string.

For the middle size string the pen will be 10 cm (4 in.) from the thumbtack, and for the longest piece of string the pen will be 15 cm (6 in.) from the thumbtack.

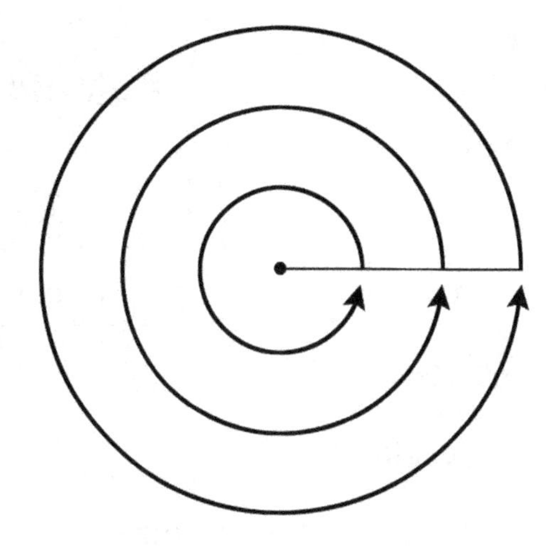

Results

Using the formulas provided in the table on the next page, calculate the tangential speed of each circle.

❶ With the ruler, measure the radius of each circle and write these numbers in the table.

❷ Using the equation in the calculation box, calculate the circumference for each circle.

❸ Calculate the tangential speed for each circle, assuming a rotational speed of 1 RPM (revolution per minute).

Calculating Tangential Speed

	①	②	③
String Length	5 cm	10 cm	15 cm
Circle:			
Radius	______	______	______

Calculate the Circumference [∏ (pi) = 3.14]

①	②	③
c = 2∏r	**c = 2∏r**	**c = 2∏r**
c = 2∏r · ______	c = 2∏r · ______	c = 2∏r · ______
c = ______	c = ______	c = ______

Circumference ______ ______ ______

Calculate the tangential speed for one revolution (1 RPM)

Tangential speed = distance traveled/time

Note: Distance traveled is one revolution — the circumference of the circle (c).

Time (t) equals one minute for this problem.

Tangential speed (S_T) = c/t

Calculation

①	②	③
$S_T = c/t$	$S_T = c/t$	$S_T = c/t$
$S_T =$ ______ /t	$S_T =$ ______ /t	$S_T =$ ______ /t

Tangential Speed ______ ______ ______

III. Conclusions

Which circle has the highest tangential speed? ______________________

Which circle has the lowest tangential speed? ______________________

What conclusions can you draw from your observations?

____

____

____

____

____

____

____

____

____

____

____

____

____

____

____

____

____

IV. Why?

One of the great things about physics is that mathematics will often confirm your experience. If you hit the gas pedal in a car, you will feel the acceleration, and you can show this with mathematics. If you suddenly stop by hitting the brake, you will feel the change in momentum, and you can also show this mathematically.

By doing this experiment you can experience how tangential speeds are greater on the outside of a circle than towards the center. Another way to make this observation is to put your finger on a rotating disk and note how your finger moves faster when it is near the outside of the disk than when it is near the center. If you sit first on the outside of a merry-go-round that is spinning and then sit near the center, you can feel the difference in tangential speed.

In this experiment you calculated the difference in tangential speed between three circles of different sizes. The larger circles have a greater radius than the smaller circles, and thus they have a larger circumference, or distance around the circle. If the rotational speed is the same for all three circles, you can see that the tangential speed increases as the size of the circle increases. This is what you experience if you sit at different distances from the center of a merry-go-round.

So the next time you go to the amusement park, if you want to spin faster or spin slower on the Spinning Top Ride, you'll know whether to choose a spot closer to or farther away from the center!

V. Just For Fun

What objects in motion have you observed that have tangential speed? List them on the following page.

How would you measure the tangential speed of one of these objects? Record your ideas about measuring the tangential speed of this object.

Tangential Speed

Moving objects that have tangential speed

Ideas for measuring tangential speed of ____________________

Experiment 9

Power Pennies

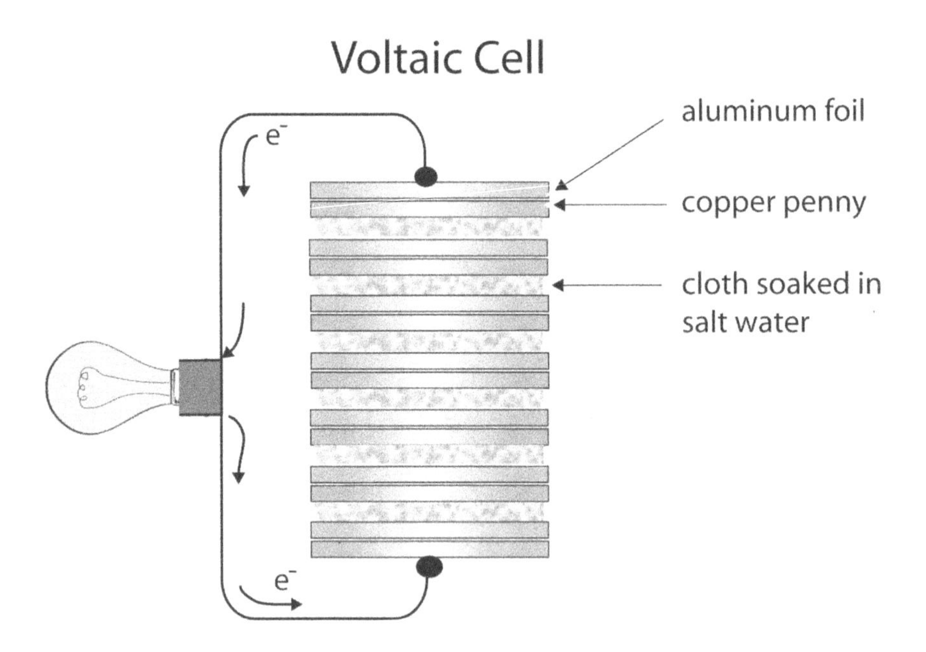

Introduction

Explore chemical energy by constructing a voltaic battery.

I. Think About It

❶ How do you think a voltaic battery works?

____

____

____

____

____

❷ What do you think would happen if you put the parts of a layer of a voltaic battery in a different order? Why?

____

____

____

____

____

❸ How can you use the chemical potential energy in a battery? Why?

____

____

____

____

____

❹ How do you think the invention of the battery changed the way people live?

❺ How many things can you think of that are powered by batteries?

❻ What do you think the difference is between mechanical energy and chemical potential energy?

II. Experiment 9: Power Pennies

Date ___________

Objective

____

____

Hypothesis

____

____

____

Materials

10-20 copper pennies
aluminum foil
paper towels
salt water: 30-45 ml (2-3 Tbsp.) salt per 240 ml (1 cup) water
voltmeter
2 plastic-coated copper wires, each 10-15 cm (4"-6") long
duct tape (or other strong tape)
scissors
wire cutters
fine steel wool

Optional

wire stripping tool

EXPERIMENT

❶ Scrub the pennies with steel wool.

❷ Cut out up to 20 penny-size circles from the aluminum foil and from a paper towel. It is important that the cutouts be very close to the size of a penny.

❸ Soak the paper towel circles in the salt water.

❹ Strip the plastic coating off both ends of one piece of wire. Use wire cutters to carefully cut through the plastic without cutting the metal wire or use a wire stripper. Use a small piece of tape to fasten one end of the wire with exposed metal touching the penny.

❺ Strip the plastic from the ends of the second piece of wire. Tape the exposed metal on one end of the wire to a piece of aluminum foil.

❻ Begin stacking the pieces. Place the circle of aluminum foil on a firm surface with the attached wire touching the surface. Put one of the wet paper towel circles on top of the aluminum foil. The paper towel piece should be wet but not dripping. On top of the paper towel, place the penny that has the wire taped to it. It should look like this:

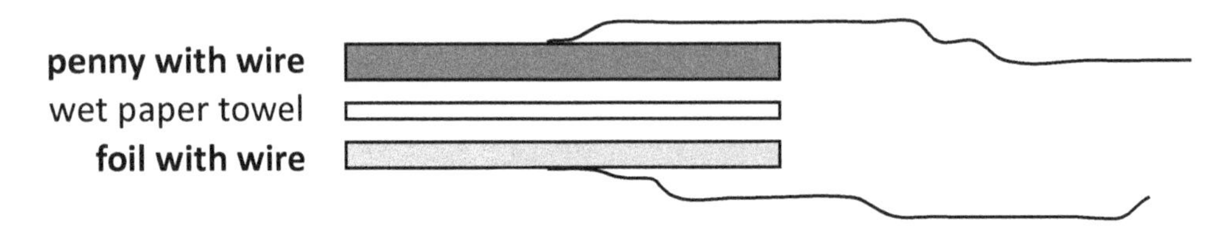

❼ Take the battery wires and connect them to the leads (wires) of the voltmeter. Switch the voltmeter to "voltage" and in the *Results* section record the number it shows. This is the amount of voltage the single layer battery produces.

❽ Add another "cell" to the battery in between the penny with the wire and the foil with the wire, which are the "ends" of the battery. A cell is a penny layer, a paper layer, and a foil layer.

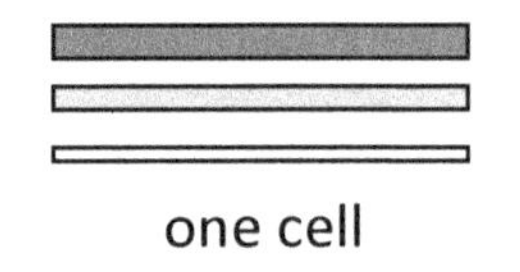

The battery now has two cells. It should look like the following:

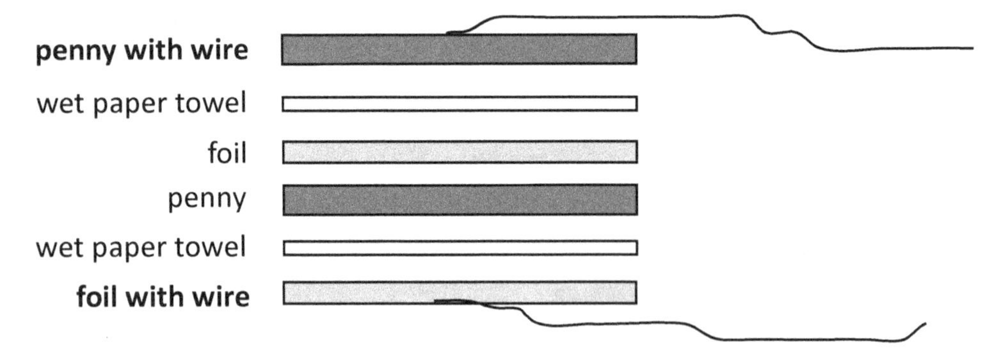

Check the voltage and record the amount in the *Results* section.

❾ Continue adding cells made of foil, wet paper towel, and pennies. Check and record the voltage after each new cell is added. Add as many cells as possible. If your results seem inconsistent, check the construction of your battery.

Results

Number of Cells	Voltage
1	
2	
3	
4	
5	

Plot your data. Make a graph with "Voltage" on the x-axis and "Number of Cells" on the y-axis.

Discuss your data. What can you observe about the data? Discuss any possible sources of error. Evaluate what worked and what didn't work.

III. Conclusions

What conclusions can you draw from your observations?

IV. Why?

In this experiment you made a battery based on Alessandro Volta's discoveries. It is thought that Volta was one of the first to observe the principles of electrochemistry, which is the scientific discipline that studies how electricity produces chemical changes and how chemical changes produce electricity.

In the late 1700s Luigi Galvani, an Italian physician, found that the muscle in a dissected frog's leg would twitch when the scalpel he was using touched a nerve while the leg was being held by a copper hook. Galvani felt that he had discovered the effects of "animal electricity," or electrical force stored in the tissues of animals. The Italian scientist Alessandro Volta reviewed Galvani's experiments and thought the result might actually be caused by the contact of the two metals with the moist frog tissues. Volta's experimentation led to the invention of the first electric battery, sometimes called the voltaic pile. Similar to the battery you constructed, Volta's battery was made of alternating discs of leather or paper soaked in a salt solution (or vinegar) and layered between copper and zinc discs. Volta found that an electric current could be generated in this way.

A battery is specifically designed to convert chemical potential energy into electrical energy by means of chemical reactions. In each cell of your battery, the aluminum foil has a chemical reaction with the salt water which causes the aluminum to give up electrons, thus becoming positively charged. The copper penny also has a chemical reaction with the salt water and takes on electrons, becoming negatively charged.

Salt water is a conductor and electrons can be conducted through the salt water from the aluminum foil to the copper penny. In the adjoining cell, the copper penny is touching the foil . Electrons move from the negatively charged copper to the positively charged aluminum which then reacts with the salt water and the process continues through all the cells. When you attach a wire to the aluminum foil at one end of the battery and a second wire to the penny at the other end of the battery and then attach the ends of these wires to a voltage meter, you have made an electrical circuit that allows the electrons to flow from the aluminum foil at one end of the battery through all the cells to the penny at the other end. The electrons then flow from the penny through the wire to the voltage meter and through the voltage meter to the aluminum foil, and the process continues.

A device that can run on the amount of electric current generated by the battery, such as an LED light, can be inserted in place of the voltage meter and will be powered by the electrical current from the battery. The difference in electrical potential (voltage) between the positively charged aluminum and the negatively charged penny creates the flow of electric force.

V. Just For Fun

Steel Wool Meets Battery

What do you think will happen when steel wool meets a battery?

Materials

9 volt battery
fine steel wool, plain (no soap)
ovenproof pan or dish
heatproof pad or surface

This experiment is best done outside. Have an adult help you.

1. Take a piece of steel wool and spread it out with your fingers until it is a loose ball.
2. Place the steel wool ball in an ovenproof pan that is on a heatproof pad or surface.
3. Hold the 9 volt battery at the bottom and quickly and carefully touch the terminals to the steel wool. What happens?
4. Record your observations.

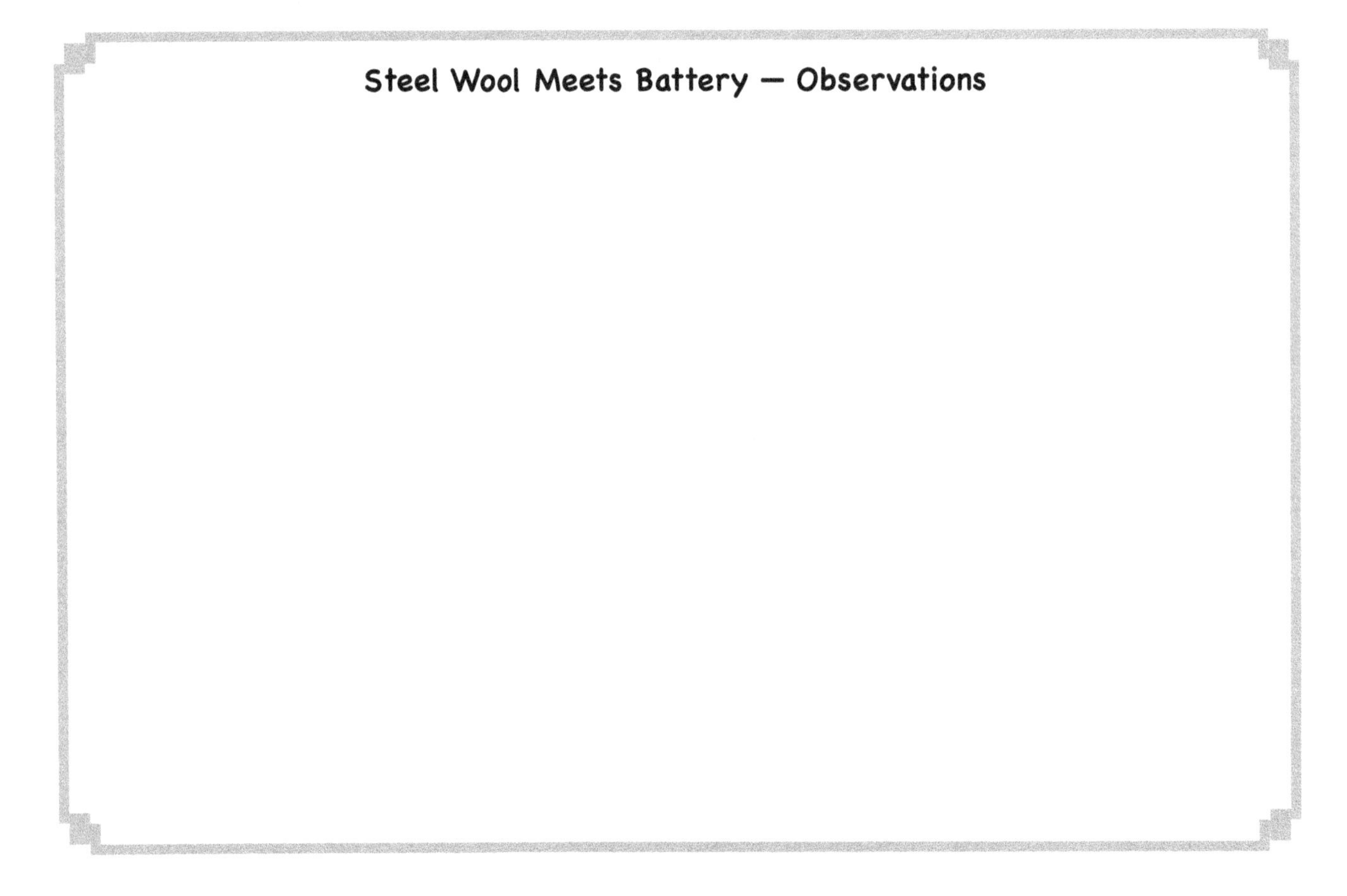

Why did what happened happen?

In the space below write your conclusions, then send a summary to us at: office@gravitaspublications.com.

If you're right, we'll send you a surprise gift!

Experiment 10

Charge It!

Introduction

Build your own instrument to detect electric charge—an electroscope!

I. Think About It

❶ What are the parts of an atom that have a charge and what charge do you think they carry? How do you think electric charge is important to an atom?

____

____

____

____

____

❷ What do you think happens when a positive charge meets a negative charge? Why?

____

____

____

____

____

❸ What do you think happens when a negative charge meets another negative charge? Why?

____

____

____

____

____

❹ What do you think would happen if all the parts of an atom were positively charged? Why?

____

____

____

____

____

____

❺ What do you think has happened when an object goes from having no charge to having a negative charge? From no charge to a positive charge? Why?

____

____

____

____

____

____

❻ How would you explain electrical force to someone who doesn't know much about physics?

____

____

____

____

____

____

II. Experiment 10: Charge It!

Date ___________

Objective

____

____

Hypothesis

____

____

____

Materials

small glass jar with lid
aluminum foil
paperclip
duct tape (or other strong tape)
plastic or rubber rod (or balloon)
silk fabric (or your hair)
scissors
ruler
awl or other tool to make a hole

EXPERIMENT

❶ Cut two narrow strips of aluminum foil of equal length (about 2.5 cm [1 inch] long).

❷ Poke a small hole in the center of the lid of the glass jar.

❸ Bend open a paperclip to make a right angle from the outer loop of wire and a small hook from the inner loop. (See illustration on the right.)

❹ Push the straight part of the paperclip through the small hole in the jar lid starting from the bottom side of the lid. Secure the paper clip to the outside of the lid with strong tape, leaving the end of the paper clip exposed. (See illustration on next page.)

5. Hang the two strips of aluminum foil from the hook that is on the underside of the jar lid. Place the lid on the jar with the aluminum foil hanging from the hook inside the jar.

   You now have an electroscope.

6. Take the plastic or rubber rod and rub it with the silk fabric, or take the balloon and rub it in your hair or on the cat.

7. Gently touch the rod or the balloon to the end of the paper clip that is on the outside of the jar lid.

8. Observe the two pieces of aluminum foil and record your results.

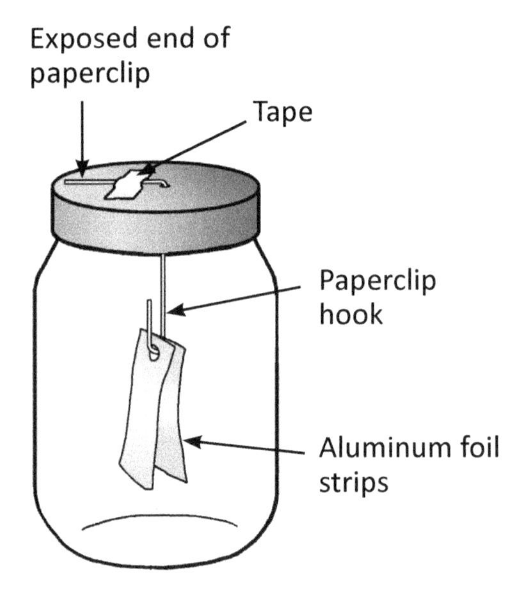

Results

III. Conclusions

What conclusions can you draw from your observations?

IV. Why?

Static electricity is generated when an object like a balloon is rubbed on hair. Friction causes electrons to move from the hair to the balloon or the silk fabric to the rod, causing the rod or the balloon to be negatively charged. Electrons have a negative charge, so an object that gains electrons will become negatively charged.

The electroscope works with static electricity and the fact that like charges repel. In a charged object the like charges always move as far away from each other as they can. When you touch the paperclip in the electroscope with a charged object such as the balloon or plastic rod, the electrons move as far apart as possible, spreading from the rod along the paperclip and then along the aluminum strips. Both of the foil strips then have negative charge, so they repel and move away from each other.

After awhile the charge leaks away, and the foil strips come back together. The more charge, the more strongly the strips repel; therefore, you can tell how strongly charged the rod was when you started.

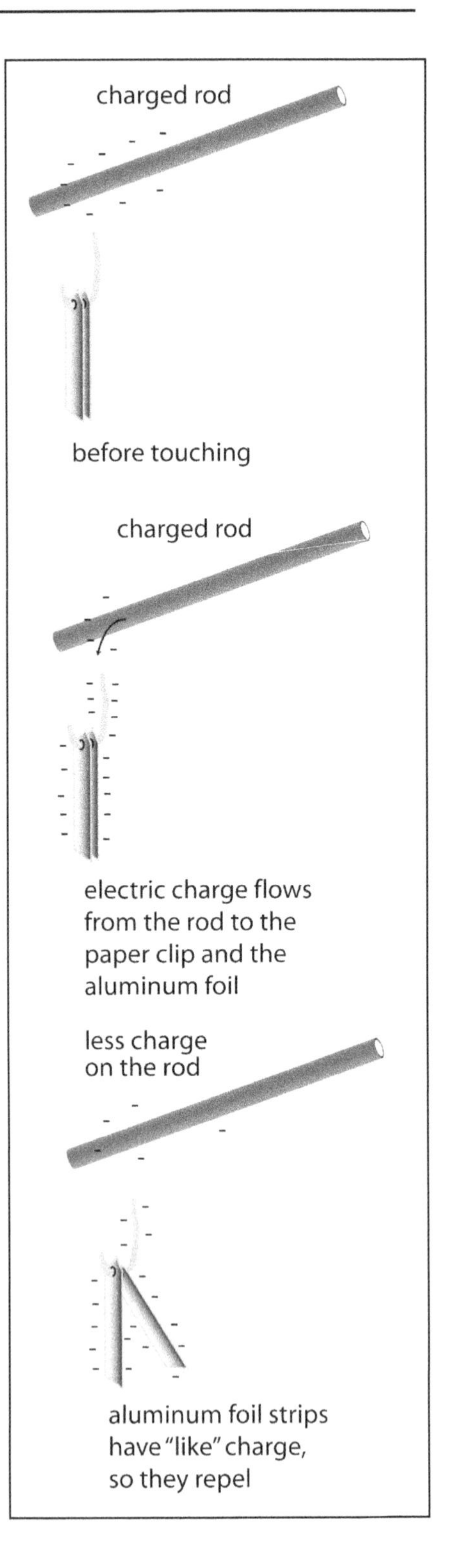

V. Just For Fun

More static electricity experiments!

Here are some quick experiments that show static electricity in action. For these to work well, you need to be in a room with low humidity. For each experiment, charge the straw or comb by rubbing it with a paper tissue or a cloth made of silk or wool. You may want to have a helper when you're charging two objects so you don't discharge one of them by touching it while you are charging the other.

Space is provided for you to record your observations.

Materials

several thin, bendable plastic straws
paper tissues (Kleenex) or cloth made of silk or wool
small piece of paper
small piece of aluminum foil
scissors
1 or more books
1 or 2 plastic combs
plastic cup
shallow bowl or a plate

- Cut some very small pieces of paper and aluminum foil. What do you think will happen when you put a charged straw near the pieces? How close do you think the straw will have to get for something to happen?

- Do you think you can use a charged straw to turn a page in a book?

- Turn on a faucet so a thin stream of water is coming out. Charge a plastic comb. What do you think will happen as you move the charged comb slowly toward the water? What do you think will happen if you move the comb around or wiggle it? Do you think anything different will happen if you use two charged combs?

  Charge a straw. Repeat the experiment. Do you notice any differences between how the water reacts to the straw and to the comb? To two charged straws? A comb and a straw?

- Bend two straws slightly, then charge them. Hold one in each hand by the short part so the long parts are parallel and upright. What do you think will happen as you move them close together?

- Charge two straws. Hold them so they are parallel to the floor and aligned one above the other and with your hands one above the other. What do you think will happen as you move the straws closer together? (Hold the top straw loosely—just enough to keep it aligned over the bottom one and not swinging from side to side.)
- Take a shallow bowl or plate and put just enough water in it that a plastic cup will float. Charge one end of a straw and place the straw on top of the cup (across the rim). What do you think will happen if you charge another straw and bring it close to the straw that's on the cup? What will happen if you move the second straw around? Do you think if you used a charged comb, you would get different results? Two straws? Two combs? A straw and a comb? What if you charged the whole straw that's on the cup instead of just one end?
- What other static electricity experiments can you make up?

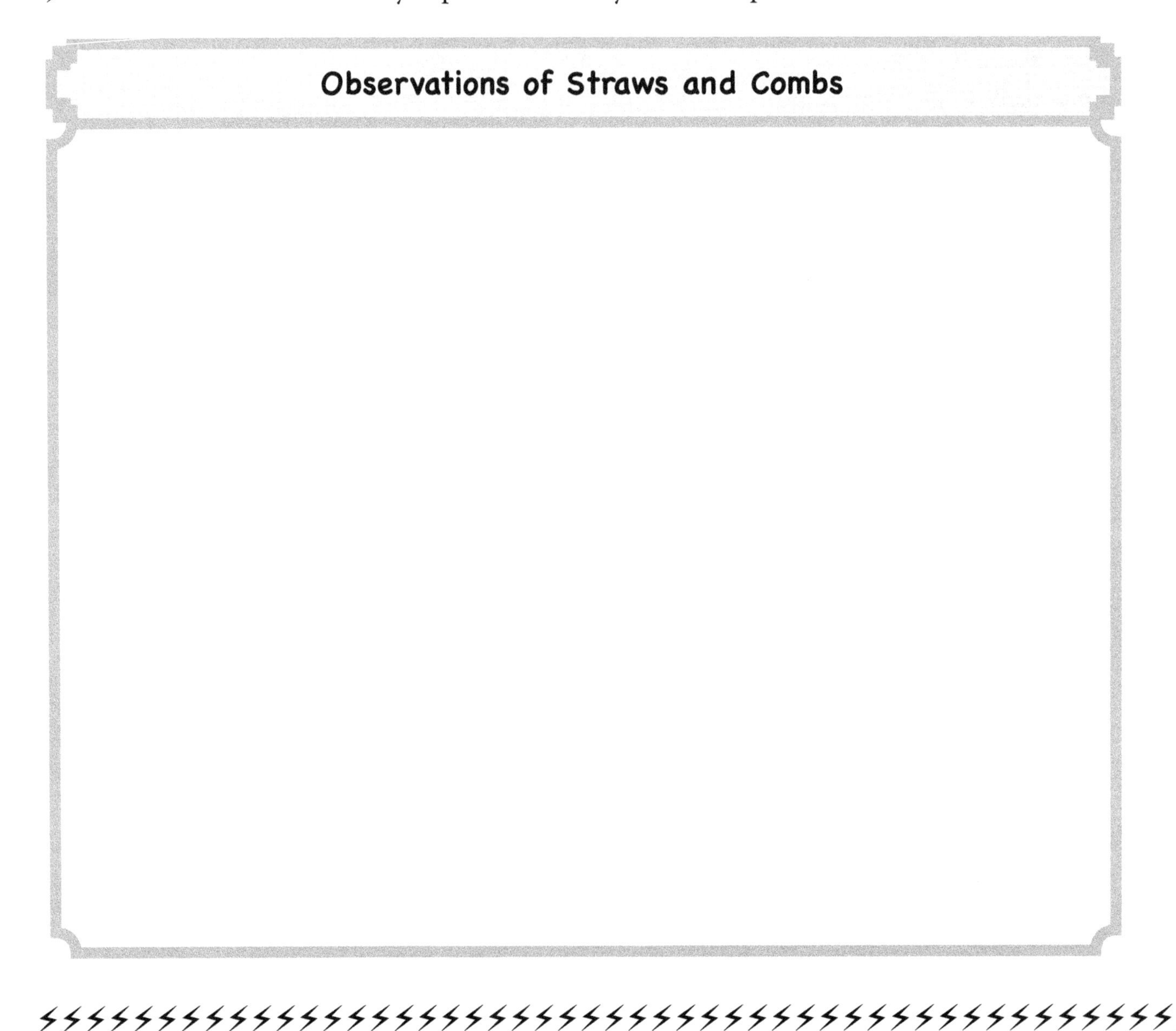

More Observations of Straws and Combs

Experiment 11

Circuits and Ohm's Law

Introduction

Explore electric circuits.

I. Think About It

❶ How do you think an electric circuit works?

____

____

____

____

____

❷ What do you think is needed for an electric current to work?

____

____

____

____

____

❸ How would you describe Ohm's Law in your own words?

____

____

____

____

____

❹ How much voltage do you think it would take to light a small light bulb? Why?

❺ How much current do you think it would take to light a small light bulb? Why?

❻ What do you think would happen if you put a resistor in your circuit? Would the light bulb still light? Why or why not?

II. Experiment 11: Circuits and Ohm's Law

Date ____________

Objective ____

____

Hypothesis ____

____

____

Materials	**Symbol**
(2) D cell batteries and battery holder	battery
(1) 3.7 volt light bulb and socket	light bulb
(1) switch	switch
(4) alligator clip connectors	wire
(2) 5 ohm, ¼ watt resistors	resistor
(1) DC motor with propeller	motor

EXPERIMENT

Part I: Building a simple circuit

❶ Using the symbols in the *Materials* section, draw a simple circuit. Include the battery source, the light bulb, and the switch.

❷ Build a circuit according to your diagram and close and open the switch. Record your observations below.

❸ Repeat Step ❶ and include the DC motor in your drawing of the circuit.

❹ Build your circuit, close and open the switch, and record your observations below.

Part II: Testing Ohm's Law

❶ Build the following circuit. Close the switch and note the brightness of the bulb. Record your observations.

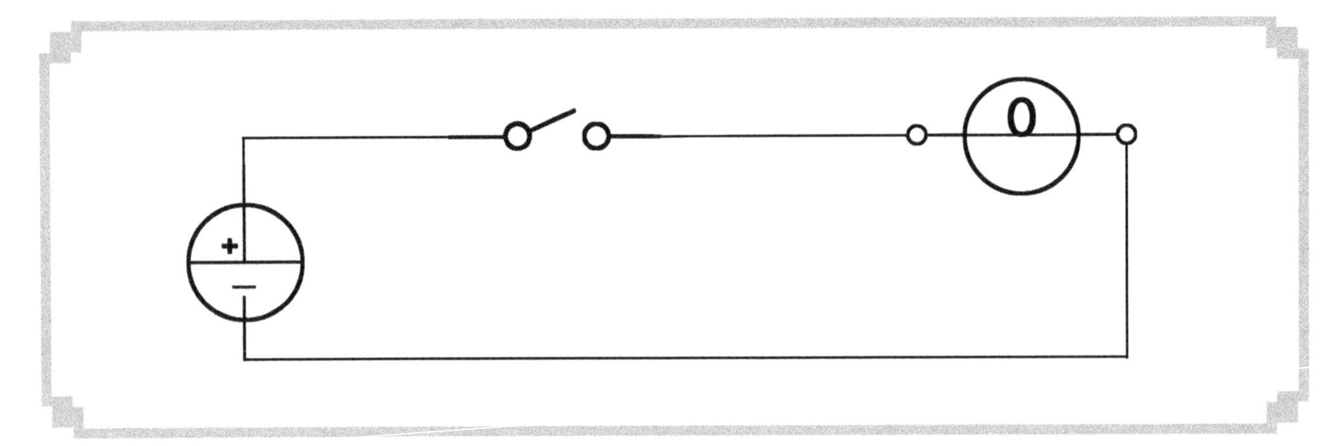

❷ Open the switch and insert a resistor between the light bulb and the battery.
Draw your circuit.
Close the switch and observe the brightness of the light bulb.
Record your observations.

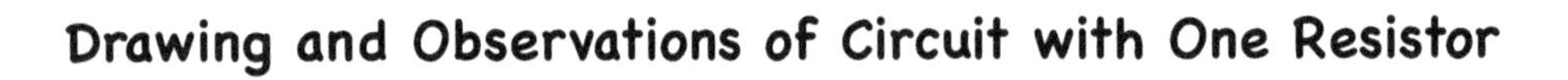

❸ Open the switch and insert a second resistor next to the first one. You may need to wrap or bend the end wires of the resistors together to make a good connection.
Draw your circuit.
Close the switch and observe the brightness of the light bulb.
Record your observations.

Drawing and Observations of Circuit with Two Resistors

III. Conclusions

Review your data. What conclusions can you draw from your observations? What does your experiment tell you about Ohm's Law?

IV. Why?

An electrical circuit is made when a power source is connected to a conducting wire and some type of load. A load is an electrical component or part of a circuit that draws, or consumes, electric power. A load could be a resistor, a motor, an electrical appliance, or a light bulb. In this experiment you created a simple circuit with all the elements (battery, conducting wire, and resistor) connected in a *serial circuit*. A serial circuit gets its name because all the components are connected in a *series* (one after another).

The electrical current in this experiment is created by chemical reactions inside the battery. When the switch is down, the circuit is *closed*, and electric current will flow from the battery through the conducting wires and through the filament in the light bulb, causing the bulb to light up. When the circuit is *open* (switch is open), the path of the electrical current through the wire is interrupted so no current can flow—no electrons can travel through the wire to illuminate the light bulb.

When a resistor is added to the circuit, the intensity of the light bulb is diminished because the resistor reduces the amount of electric current that can reach the bulb. Using Ohm's Law, you can see that as the resistance is increased, the current decreases.

Ohm's Law:

$$I = \frac{V}{R}$$

You can see from this equation, that as R (resistance) increases, I (current) decreases.

If two resistors are added to the circuit, there is enough resistance reducing the flow of electrons that the light bulb will not illuminate because there is not enough electric current getting to it.

V. Just For Fun

Create a parallel circuit with two light bulbs. Draw your circuit and record your observations.

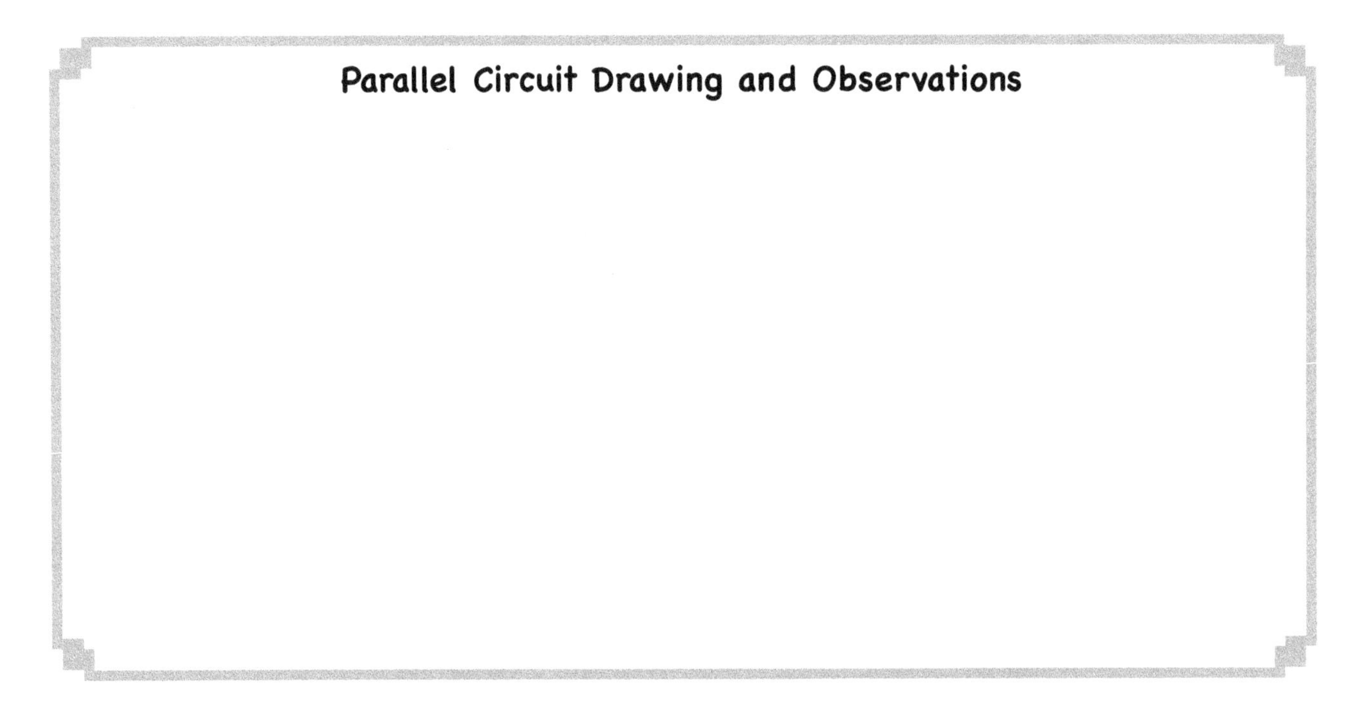

Using the materials you have, build more circuits with your own variations. Draw the circuit and record your results for each variation. When you have finished the experiment, review your data and see what conclusions you can draw.

More Circuit Drawings and Observations

Even More Circuit Drawings and Observations

Conclusions

Review your data. What conclusions can you draw from your observations? What worked and didn't work? Why? What more did you learn about electrical circuits and Ohm's Law? What did you learn about serial and parallel circuits?

Experiment 12

Wrap It Up!

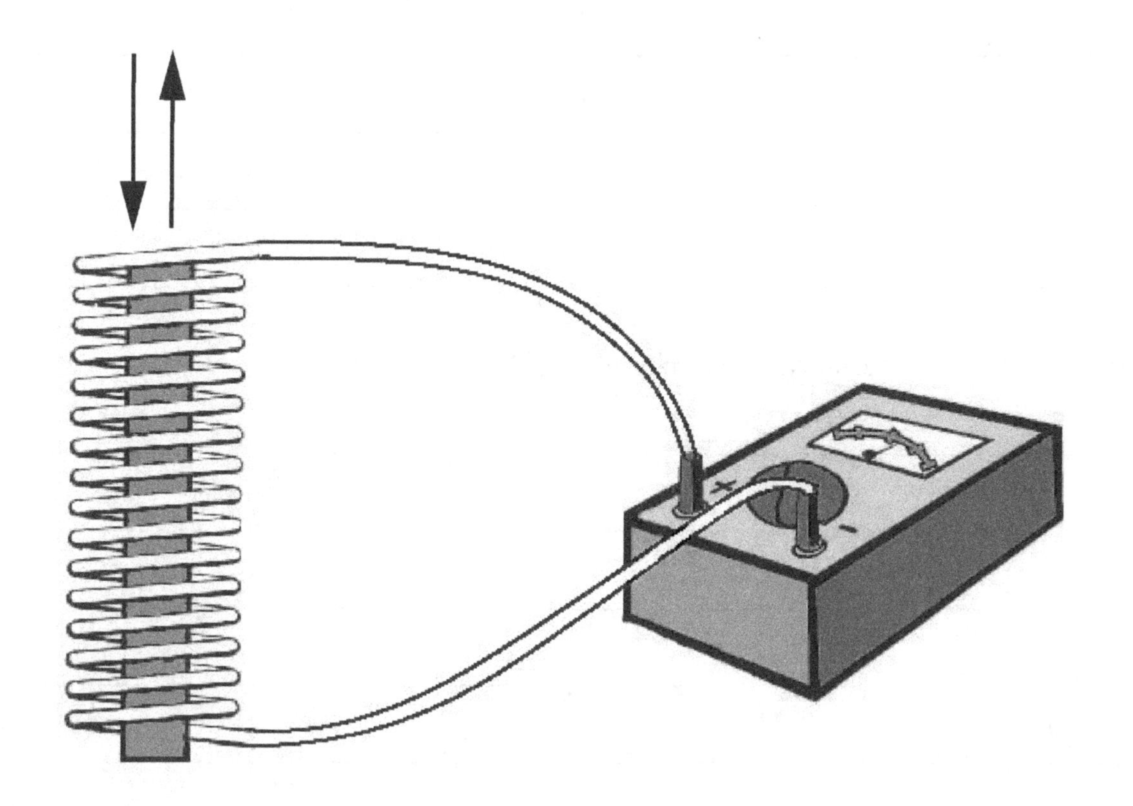

Introduction

Do you think you can make a magnet with a battery and some wire? Find out!

I. Think About It

❶ How would you find out if an object contains magnetic force?

❷ Do you think any material can be a magnet? Why or why not?

❸ What do you think would happen if you cut a magnet into little pieces? Why?

❹ What happens if you place two magnets with their north poles together? Why?

❺ Do you think magnetic force can be useful? Why or why not?

❻ Do you think an electromagnet might have any advantages over a bar magnet? Why or why not?

II. Experiment 12: Wrap It Up!

Date ___________

Objective

____

____

Hypothesis

____

____

____

Materials

metal rod (a large nail or unmagnetized screwdriver can be used)
electrical wire
10-20 paperclips
6v or larger battery
electrical tape or 2 alligator clips
scissors
wire cutters
Optional: wire stripping tool

EXPERIMENT

❶ Cut the electrical wire so that it is .3-.6 meter (1-2 feet) long.

❷ Trim the plastic coating off the ends of the wire so that there is about 6 mm (1/4 inch) of exposed metal on each end of the wire.

❸ Tape one end of the wire to the positive (+) terminal of the battery. (Alligator clips may be used in place of tape.)

❹ Tape (or clip) the other end of the wire to the negative (-) terminal of the battery.

❺ Take the metal rod and touch it to the paperclips. Record your results in the chart in the *Results* section.

❻ Coil the wire around the metal rod a few times. Both ends of the wire must remain hooked to the battery or be reattached if they come off.

❼ Touch the metal rod to the paperclips. Count the coils and record your results.

8. Wrap another 1 to 5 coils around the metal rod.

9. Touch the end of the metal rod to the paperclips. Record the number of coils and how many paper clips were picked up.

10. Continue adding coils to the metal rod and counting the number of paperclips that can be picked up. Record the results each time you increase the number of coils. When you are finished, make a graph of your data.

Results

Number of Coils	Number of Paperclips

Graph your results

III. Conclusions

Review your data. What conclusions can you draw from your observations?

IV. Why?

An electromagnet is created when an electric current flows through a wire and causes a magnetic field that circles the wire. The more electric current that passes through the wire, the stronger the electromagnet will be — when there is more electric current, there is a stronger magnetic field.

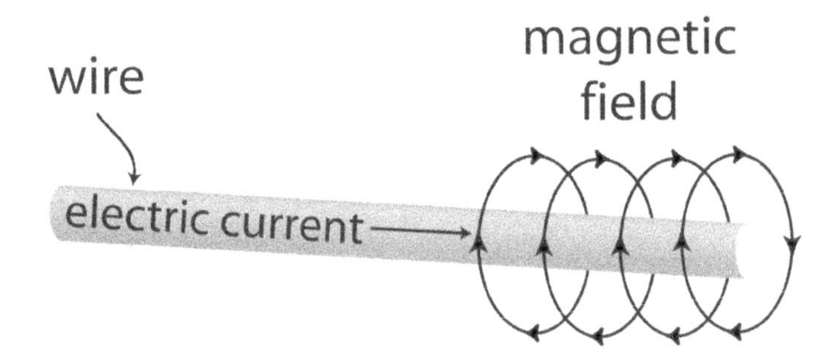

If the wire is wound into a coil, it can behave much like a bar magnet when the electric current is passed through it. The fields from each part of the coil add up to create a magnetic field that looks much like the field of a bar magnet.

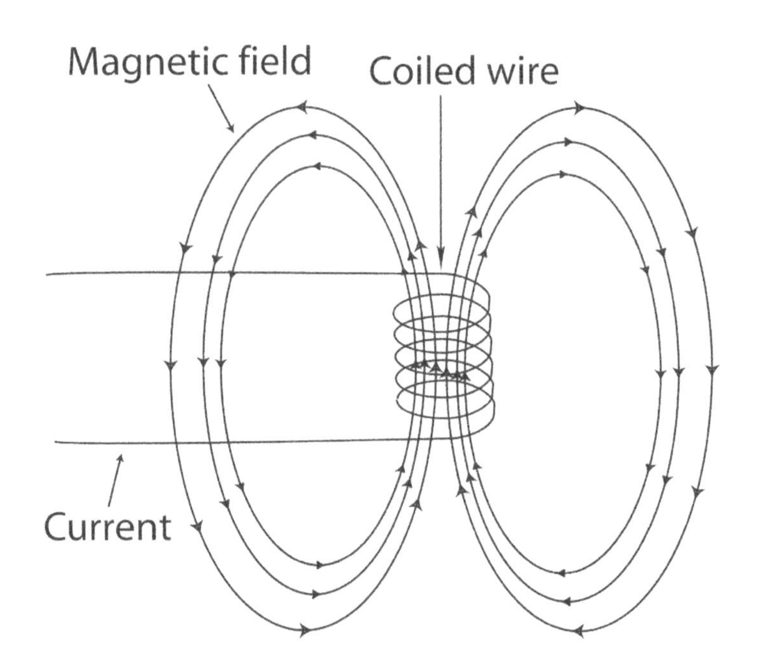

In this experiment you observed that the electromagnet you created became stronger (picked up more paperclips) when you increased the number of loops around the metal rod and thus increased the strength of the magnetic field. Another way to increase the strength of the magnetic field would be to use a larger battery. You would notice that as the current in the wire increased with the stronger battery, a stronger electromagnet would be produced.

V. Just For Fun

Seeing the Magnetic Field

Materials

bar magnet
plastic baggie
small flat-bottomed, clear plastic container with lid [about 5 cm x 8 cm x 1.5 cm (2" x 3" x 1/2") is a good size]
clear Karo syrup
spoon
2 pencils or other props
electromagnet from Wrap It Up! experiment

Experiment

❶ Take the bar magnet and place it in a plastic baggie. Take the bar magnet and the plastic container outside to collect iron filings from your yard or another place where there is exposed dirt. Hold the baggie containing the magnet and swirl it in the dirt. Small iron filings should collect on the outside of the baggie around the magnet.

❷ Carefully place the baggie in the plastic container and remove the magnet. The iron filings should fall off the baggie and into the box.

❸ Repeat Steps ❶-❷ several times until you have enough iron filings to thinly cover the bottom of the box.

❹ Pour enough Karo syrup into the box to just cover the iron filings and gently stir.

❺ Carefully place the plastic container on top of the electromagnet. Put props under the the container to lift it slightly above the electromagnet and keep it flat.

❻ Turn on the electromagnet and observe what happens over the next several minutes.

❼ Record your conclusions in the following space.

Conclusions

What conclusions can you draw from your observations?

More REAL SCIENCE-4-KIDS Books
by Rebecca W. Keller, PhD

Building Blocks Series yearlong study program — each Student Textbook has accompanying Laboratory Notebook, Teacher's Manual, Lesson Plan, Study Notebook, Quizzes, and Graphics Package

Exploring the Building Blocks of Science Book K (Activity Book)
Exploring the Building Blocks of Science Book 1
Exploring the Building Blocks of Science Book 2
Exploring the Building Blocks of Science Book 3
Exploring the Building Blocks of Science Book 4
Exploring the Building Blocks of Science Book 5
Exploring the Building Blocks of Science Book 6
Exploring the Building Blocks of Science Book 7
Exploring the Building Blocks of Science Book 8

Focus Series unit study program — each title has a Student Textbook with accompanying Laboratory Notebook, Teacher's Manual, Lesson Plan, Study Notebook, Quizzes, and Graphics Package

Focus On Elementary Chemistry
Focus On Elementary Biology
Focus On Elementary Physics
Focus On Elementary Geology
Focus On Elementary Astronomy

Focus On Middle School Chemistry
Focus On Middle School Biology
Focus On Middle School Physics
Focus On Middle School Geology
Focus On Middle School Astronomy

Focus On High School Chemistry

Super Simple Science Experiments

21 Super Simple Chemistry Experiments
21 Super Simple Biology Experiments
21 Super Simple Physics Experiments
21 Super Simple Geology Experiments
21 Super Simple Astronomy Experiments
101 Super Simple Science Experiments

Note: A few titles may still be in production.

Gravitas Publications Inc.
www.gravitaspublications.com
www.realscience4kids.com

CPSIA information can be obtained
at www.ICGtesting.com
Printed in the USA
LVHW061221170319
610948LV00027B/1729/P